SEE HOW IT'S MADE

BUILDING A SPACECRAFT

BY TYLER OMOTH

CONTENT CONSULTANT
RYAN P. STARKEY, PHD
DEPARTMENT OF AEROSPACE
ENGINEERING SCIENCES
UNIVERSITY OF COLORADO, BOULDER

READING CONSULTANT
BARBARA J. FOX
PROFESSOR EMERITA
NORTH CAROLINA STATE UNIVERSITY

Blazers Books are published by Capstone Press,
1710 Roe Crest Drive, North Mankato, Minnesota 56003
www.capstonepub.com

Library of Congress Cataloging-in-Publication Data
Omoth, Tyler.
Building a spacecraft / by Tyler Omoth.
pages cm — (Blazers. See how it's made)
Summary: "Describes the process of building a spacecraft"— Provided by publisher.
Audience: Age 9.
Audience: Grades 4 to 6.
Includes bibliographical references and index.
ISBN 978-1-4765-3979-9 (library binding)
ISBN 978-1-4765-5119-7 (paperback)
ISBN 978-1-4765-5960-5 (ebook pdf)
1. Space vehicles—Design and construction—Juvenile literature. I. Title.
TL875.O46 2014
629.47—dc23 2013032489

Editorial Credits
Mandy Robbins, editor; Kyle Grenz, designer; Kathy McColley, production specialist

Photo Credits
Corbis: Roger Ressmeyer, 9; NASA, 1, 5, 6, 13, 14, 17, 18-19, 21, 23, 25, 26, 29, cover (satellite); NASA EDGE: Ron Beard, 10-11; Shutterstock: Nikkolia, cover, 1 (inset laser), Saulius L, cover (background), sparkdesign, throughout (background)

Printed in the United States of America in Stevens Point, Wisconson.
092013 007768WZS14

TABLE OF CONTENTS

INTO THE GREAT UNKNOWN

Exploring space is full of danger and excitement. Astronauts face darkness and extreme temperatures in space. They need spacecraft that are strong and dependable.

FACT

Astronauts can grow up to 2 inches (5 centimeters) in space. The force of gravity is weaker in space, so astronauts' spines stretch out more, increasing their height.

astronaut—a person who is trained to live and work in space

gravity—a force of attraction between two objects; for example, the sun's gravity holds Earth and the other planets in orbit around it

United States

INTERNATIONAL
SPACE STATION

FACT
The *International Space Station (ISS)* is about the size of a football field.

There are many types of spacecraft. Space stations and satellites circle the Earth. Other spacecraft include rockets, space shuttles, and probes. Each spacecraft is built for a special purpose.

space station—a spacecraft that circles the Earth and is large enough to house a crew for long periods of time

satellite—an object that circles a planet; some satellites that circle Earth study land and weather or help with communication

space shuttle—a spacecraft that carries astronauts who work in space and returns to Earth when its mission is complete

probe—a small vehicle used to explore objects in outer space

FIRST STEPS

Engineers use computers to create spacecraft designs and models. Computer tests help to find the best shapes for the models. The shape of a vehicle affects how it moves through space.

FACT
Building a spacecraft is a long process. The *Falcon 9* rocket took four and a half years to design and build.

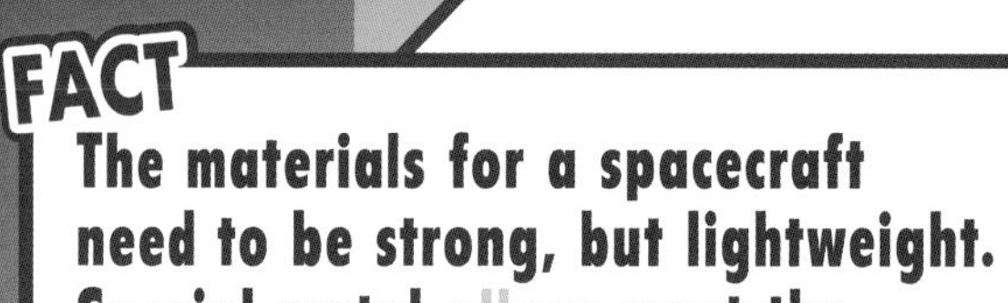

FACT

The materials for a spacecraft need to be strong, but lightweight. Special metal alloys meet the demands of space travel.

Scientists build a scale model for some spacecraft once a design is chosen. The model tests the limits of the materials and the shape of the design.

alloy—a combination of two or more metals

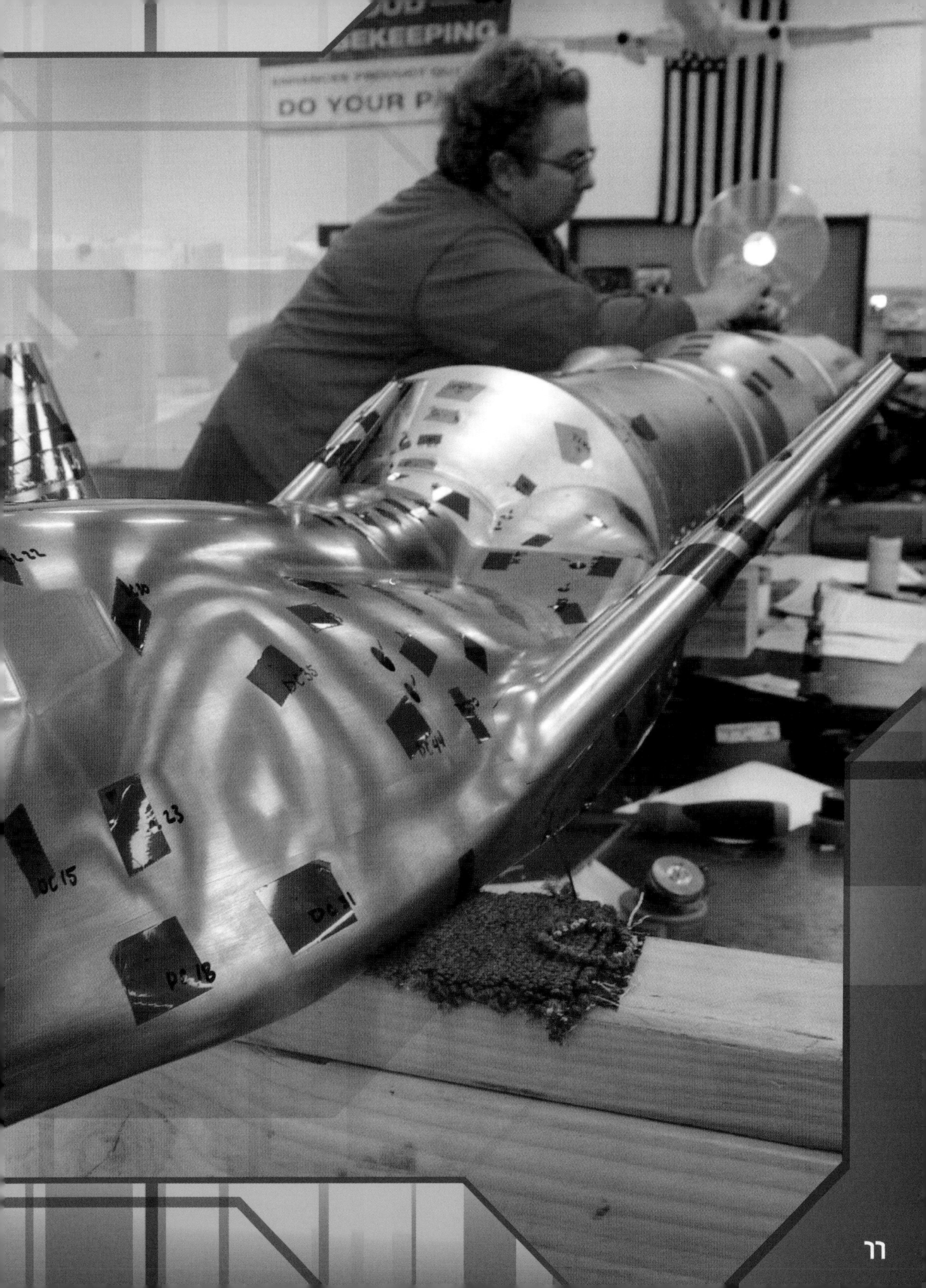
DO YOUR
DC 15
DC 18

CHAPTER 2

PUTTING IT TOGETHER

Rockets, shuttles, and space stations are made in pieces. Factories build parts of the crafts and send them to the building site. Engineers and other scientists watch over each step of the process.

▲ Crews prepare to transport a section of the *ISS* to the building site.

FUSELAGE

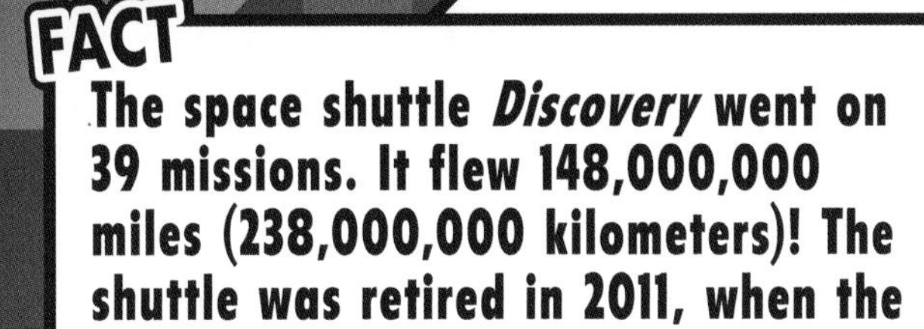

FACT

The space shuttle *Discovery* went on 39 missions. It flew 148,000,000 miles (238,000,000 kilometers)! The shuttle was retired in 2011, when the U.S. space shuttle program ended.

The fuselage of a space shuttle is built in three main sections. Aluminum sheeting covers strong steel ribs.

All parts of a spacecraft go through many tests. The outside surfaces must survive temperatures of 2,500 degrees Fahrenheit (1,371 degrees Celsius) or higher as they reenter the atmosphere.

FACT

More than 21,000 tiles cover the bottom of a space shuttle to protect it from the heat of reentry.

atmosphere—the mixture of gases that surrounds the Earth

The *Orion* manned spacecraft is tested to see if it can survive launch forces.

SECTION 352

Workers attach an antenna to a power source to make a satellite. The antenna sends and receives information to and from Earth. It uses power from a solar panel or battery. The smallest, most inexpensive satellites are made out of smartphones.

FACT
More than 2,500 satellites circle the Earth today. Some are no longer working.

antenna—a wire or dish that sends or receives radio waves
solar panel—a flat surface that collects sunlight and turns it into power
battery—a container holding chemicals that store and create electricity

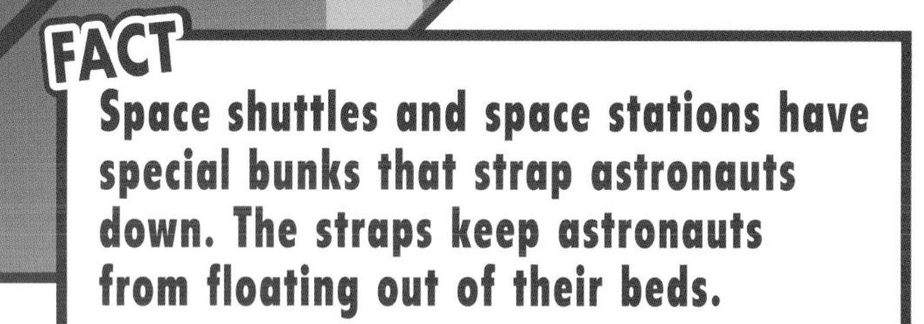
FACT
Space shuttles and space stations have special bunks that strap astronauts down. The straps keep astronauts from floating out of their beds.

People must be able to stay alive in space shuttles and space stations. Workers add life support systems to spacecraft to provide the crew with food, water, and oxygen.

Workers build the life support system for the *ISS*.

CHAPTER 3

BLAST OFF!

Building a spacecraft isn't always finished on the surface of the Earth. The *ISS* went through 40 assembly flights to complete. Rockets launched shuttles into space. There astronauts continued the building process.

FACT
The *ISS* has been visited by 204 astronauts since its launch in 2000.

Sometimes astronauts must make repairs outside the aircraft in space. This is called an Extra Vehicular Activity (EVA). During an EVA, the astronaut wears a special space suit that is connected to the spacecraft by a strap.

TRAJECTORY

It takes many people to run a space repair mission. Experts in a control center watch over the mission using computers. They help guide astronauts from Earth.

FACT

There are 20 different stations in NASA's Mission Control Center. Workers at each station monitor a different part of a space mission.

CHAPTER 4

GREAT DISCOVERIES

Building spacecraft is a long, costly process. But without them, new discoveries about space couldn't be made. Just imagine what spacecraft could help discover in the future.

FACT

The *New Horizons* space probe was launched on January 19, 2006. It is scheduled to reach Pluto in 2015.

▲ the *ISS* floating over Earth

GLOSSARY

alloy (AL-oi)—a combination of two or more metals

antenna (an-TE-nuh)—a wire or dish that sends or receives radio waves

astronaut (AS-truh-nawt)—a person who is trained to live and work in space

atmosphere (AT-muhss-fihr)—the mixture of gases that surrounds the Earth

battery (BA-tuh-ree)—a container holding chemicals that store and create electricity

engineer (en-juh-NEER)—a person who uses science and math to plan, design, or build

fuselage (FYOO-suh-lahzh)—the main body of a space shuttle

gravity (GRAV-uh-tee)—a force of attraction between two objects; for example, the sun's gravity holds Earth and the other planets in orbit around it

probe (PROHB)—a small vehicle used to explore objects in outer space

satellite (SAT-uh-lite)—an object that circles a planet; some satellites that circle Earth study land and weather or help with communication

solar panel (SOH-lur PAN-uhl)—a flat surface that collects sunlight and turns it into power

space shuttle (SPAYSS SHUHT-uhl)—a spacecraft that carries astronauts who work in space and returns to Earth when its mission is complete

space station (SPAYSS STAY-shuhn)—a spacecraft that circles Earth and can house a crew for long periods of time

READ MORE

Eason, Sarah. *How Does a Rocket Work?* How Does It Work? New York: Gareth Stevens Pub., 2010.

Jefferis, David. *Space Probes: Exploring Beyond Earth.* Exploring Our Solar System. New York: Crabtree Pub., 2009.

Parker, Steve. *Space Exploration. How It Works.* Broomall, Pa.: Mason Crest Publishers, 2011.

INTERNET SITES

FactHound offers a safe, fun way to find Internet sites related to this book. All of the sites on FactHound have been researched by our staff.

Here's all you do:

Visit *www.facthound.com*

Type in this code: 9781476539799

INDEX